Jaden Toussaint, The Greatest

Episode 4: ATTACK OF THE SWAMP THING

Published by **Plum Street Press**

ISBN 978-1-943169-18-4

Contents

Prologue vii

Chapter 1: *Camp Iwannago* 1

Chapter 2: *The Crusade* 11

Chapter 3: *The Surprise* 16

Chapter 4: *The Night Hike* 26

Chapter 5: *Swamp Thing* 35

Chapter 6: *Insomnia* 48

Chapter 7: *Tall Tales* 57

Epilogue 66

Who Made This Book? 68

HOW TO BE A SCIENTIST
ASK A QUESTION
DO BACKGROUND RESEARCH
MAKE A HYPOTHESIS
TEST YOUR HYPOTHESIS BY DOING AN EXPERIMENT
ANALYZE YOUR DATA AND DRAW A CONCLUSION
COMMUNICATE YOUR RESULTS
REPEAT!
THE BIG BOOK OF CAMPS

OCTOBER
Summer Camp

Prologue

SPEAKS KINDNESS, OOZES CONFIDENCE.

JADEN TOUSSAINT

Specializes in: Knowing Stuff. And also, ninja dancing. He's really, really good at ninja dancing.

THE FRIENDS

OWEN

Jokester and action expert.

Extreme dinosaur safari bungee jumping? Owen is your guy.

EVIE

Don't let the cuteness fool you.

This girl packs a punch. Excels at: Being in Charge.

SONJA

Cicada hunter and math whiz.

Also draws excellent rainbows.

WINSTON

Can quote stats from every World Cup final.*

*that he has been alive for

Spots hurt feelings and distracted goalies from miles away.

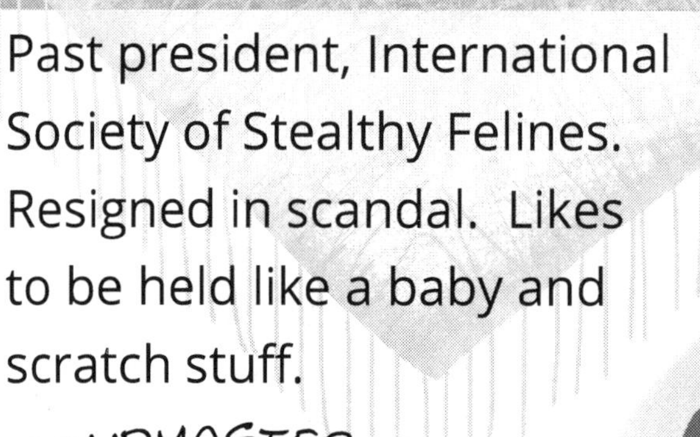

Past president, International Society of Stealthy Felines. Resigned in scandal. Likes to be held like a baby and scratch stuff.

GRANDMASTER, CAT CHESS.

GRIS GRIS

ANIMAL OF MYSTERY

Guinea Pig never has the same name two weeks in a row.

This week you can call him anything you like, as long as you pronounce it "The Artist."

THE FAM

Baba:
Tall. Competitive. Competitive about being tall. Gives great piggyback rides. Prefers to be called "baba," which means "father" in Swahili. Does not speak Swahili.

Mama:
Loving. No nonsense. Most often seen reading fantasy books or experimenting with bean desserts. Gives good hugs.

Sissy:
Reader. Writer. Animal lover. Once gave up meat for 6 months, but was broken by the smell of turkey bacon. Plans to be the first PhD chemist to star in a Broadway musical.

Chapter 1

CAMP IWANNAGO

Jaden Toussaint loved camp.

Jaden Toussaint loved camps so much that he even loved reading about them.

Every year he and Sissy would wait for the new list of summer camps to appear in the newspaper. There were so many of them!

There was zoo camp.

There was music camp.

There was Picasso camp and carnival camp.

There was circus camp and robotics camp.

There was even "So you want to be a famous animal scientist and explorer who teaches kids stuff on tv" camp!

  Pause.

Ok. He hadn't read about that one, but he could imagine it, and if it were real, it would be totally awesome.

  Unpause.

So when Jaden Toussaint heard the word "camp" coming from the living room, he raced in to join the fun.

He arrived just in time to hear Sissy say something that ended with, "Camp Planet Earth."

Camp Planet Earth? Now that could be cool. Was it a planet, or was it a camp? Or maybe it was a camp so amazingly amazing that it had evolved into its own planet? Just thinking about it made his head spin.

Then JT noticed something strange. Sissy was talking about this "Camp Planet Earth" like it was just an ordinary thing.

And normally when JT or Sissy said something exciting, Mama would give excited responses like, "Really?" or "You don't say?" And Baba would say, "Word?" or "Hunh, bruh?", which he said a lot, but he would use his super-excited voice instead of his regular voice. JT could definitely tell the difference, but Mama and Baba weren't saying any of those things.

Sissy was talking about the possible discovery of a camp-planet hybrid, and Mama and Baba were just nodding their heads. Something was definitely wrong with this picture.

It turned out that Camp Planet Earth was not a camp-planet hybrid or even a plain old planet.

"So what is it then?" JT asked.

"It's a sleep-away camp," Sissy said, pausing for effect.

"Wow!"

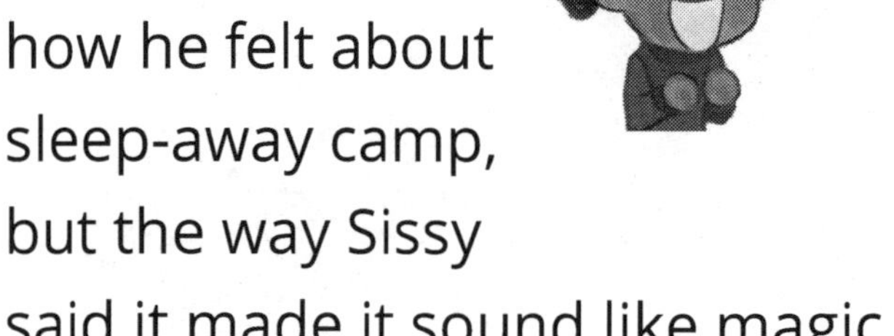

JT wasn't sure how he felt about sleep-away camp, but the way Sissy said it made it sound like magic.

"About science," she added.

"Wow!"

"And it lasts for five whole days!"

"W—"

Five whole days? JT really wasn't sure how he felt about being away from his parents for five whole days, but he and Sissy would have each other, and there would be science.... It might be fun. No. He was sure it would be fun.

"When do we leave?" he asked, wondering if his Animal Guys sleeping bag was warm enough for sleeping out in the woods for five whole days. Oh well. If it wasn't, he was sure they would let him bring some extra blankets.

"Aw, Toots," Sissy said, "you're not coming with me. Camp Planet Earth is just for 5th graders."

Just for 5th graders?

Just for 5th graders?!

He'd see about that.

Chapter 2

The Crusade

Jaden Toussaint tried everything to get to go to Camp Planet Earth.

He tried logic.

He tried reason.

He tried airing his grievances.

He even tried wearing stilts.

But on the first day of Camp Planet Earth, it was clear that Sissy was getting on the bus, and Jaden Toussaint wasn't going with her.

He was sad that he was going to miss all that superdy-awesome science, but as Sissy hugged him goodbye, he realized that the biggest thing he was going to miss was her.

What was he going to do without Sissy for five whole days? He didn't even want to think about it.

Chapter 3

The Surprise

The house felt so empty without Sissy in it.

There was no one to help him eat popcorn.

Family game night was no fun at all.

And where would he go if he had a bad dream? Sure, Mama and Baba said he could come sleep with them, but their bed was so crowded. There was barely any room for him.

Plus, neither of them were willing to talk about Toranpu cards until he fell back asleep.

With Sissy gone, Jaden Toussaint was pretty sure that he might never sleep again.

It turned out that JT wasn't the only one upset about Camp Planet Earth. Sonja's big brother had gone, too.

"Why don't we all do something together?" Mama suggested. "That might cheer you up." JT liked hanging out with Sonja, but it was going to take a lot more than a play date to brighten his mood.

Luckily, Mama was very good at mama-ing and she found something much better than a plain old play date. It was a camping trip inside the zoo, and since it was for parents and kids, kindergartners were welcome. What a relief.

Jaden Toussaint had been to the zoo lots of times. He had even been to zoo camp. But he had never even heard of people getting to sleep in the zoo. This was going to be sweeet!

With just a few phone calls, text messages, and clicks on the website, JT helped Mama arrange for Sonja and almost all the rest of his friends to sleepover in the zoo.

There would be tents and flashlights and animals. Who cared about some silly sleep-away camp when you could sleep in a tent in the zoo?

The next day, Jaden Toussaint, Mama, and Baba loaded up the car for the very short drive to the zoo. Even though the ride was short, they sang road trip songs the whole way there.

Mama had made him leave most of his animal encyclopedias at home, even after Jaden Toussaint had explained how handy they would be in the event of an animal emergency.

Turned out Mama was right. The zoo had a whole library of animal facts, and Zookeeper Jenny promised that she would leave it unlocked all night just in case they needed it.

Evie, Winston, Owen, and Sonja all set up their tents near JT's. Once the tents were put up, the air mattresses were inflated, and the marshmallows had been accounted for, everyone was ready to rough it in the wilds of the zoo.

Zookeeper Jenny told everyone the rules.

Don't feed the animals? Stay with the group? Those were excellent rules! Those were rules that made it sound like they were going to be:

They were not disappointed.

Zookeeper Jenny took them all around the zoo. They helped throw fish to the otters, pretended to be chased by an albino peacock

(everybody knows that albino peacocks have the best ninja skills), and got to go behind the scenes in the reptile house.

The reptile house made Jaden Toussaint a little sad. That was where Sissy's favorite two-headed lizard lived. But Baba told him it was okay to miss Sissy and enjoy the two-headed lizard at the same time, so he decided to do that.

It was a good thing he did, too, because otherwise he might have missed feeding meal worms to the leopard gecko, and he definitely didn't want to miss that.

After the geckos ate their dinner and the humans had eaten theirs, too, Zookeeper Jenny had one last surprise.

"Does anyone want to go on a night hike?"

Did she even have to ask?

This was officially the best trip ever.

Chapter 4

The Night Hike

The night hike was amazing.

Flashlights in hand, the whole group followed Zookeeper Jenny across the African Savannah, through the Jaguar Jungle, all the way to the Louisiana Swamp.

Not THE Louisiana swamp.
An exhibit about the Louisiana swamp.

Jaden Toussaint was brave, but walking around a real swamp at night where alligators roamed free just seemed downright foolish.

The alligators in the swamp exhibit were in the water so far below the walkway that JT knew they'd be perfectly safe, even in the dark. This was exactly the kind of adventure he didn't want to miss.

Lots of the creatures in the Louisiana Swamp are nocturnal, so the best time to see them is actually at night.

Apparently the raccoons were extra cool at night, and Zookeeper Jenny was taking the whole group to see them.

Some of the kids got distracted by baby alligators on the way. Not Jaden Toussaint.

While everyone else lagged behind, JT raced ahead. If Zookeeper Jenny was especially excited to show them the raccoons, Jaden Toussaint was especially excited to see them.

At least he was, until he realized one teensy detail:

The Loup Garou was standing right beside him.

He wasn't sure if he screamed, then ran, or ran while he was screaming. All he knew was that he had to get out of there, but when he turned to escape he was trapped by something. No. Someone.

Luckily, when he opened his eyes, that someone turned out to be Baba.

Mama and Zookeeper Jenny and the rest of the group weren't far behind.

"What's wrong, bruh?" Baba asked.
Jaden Toussaint shook his head. He was embarrassed about being frightened, and he knew he shouldn't have gone ahead.
"Did the swamp monster scare you, man?"

"It's not a swamp monster. It's the Loup Garou," Owen said with his creepiest voice.

"It's only a statue, though," Sonja added. "Nothing to be afraid of."

In his excitement to see the raccoons, JT had forgotten all about the Loup Garou. The Loup Garou was a giant wolfman who hid in the swamp and tried to scare people. Sonja was right that the one in the zoo was just a statue. He had seen it lots of times. But there was a big difference between seeing the Loup Garou during the day and seeing the Loup Garou at night.

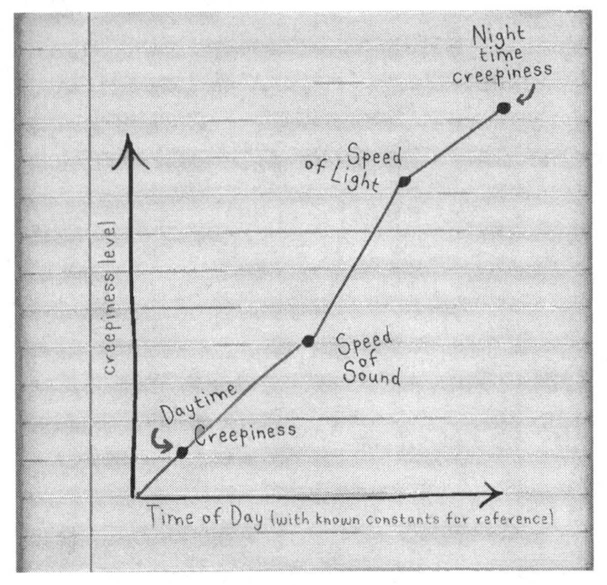

"There's nothing to fear," Winston said. "We are at harmony with nature. Besides, all the dangerous creatures in the zoo are kept behind glass."

"That's just it!" JT exclaimed. "The Loup Garou isn't kept behind a glass."

Evie jumped in. "That's because there's nothing to be afraid of, and I'll prove it. Come on," she said.

With that, Evie, Winston, Sonja, and Owen headed over to the Loup Garou with their flashlights.

Jaden Toussaint hung back. He wanted to join them, but he was just too scared.

"It's hard to be brave all the time," Mama said. "Being scared of something now just gives you a chance to be brave later on."

"I know," JT said.

  Pause.

He did know. At least he had heard Mama say it before. It was just that this was the first time it really felt real, you know?

Unpause.

Mama gave him a hug and said, "Do you want me to go with you?"

But Jaden Toussaint didn't have time to answer, because the next thing he knew, Winston, Sonja, Evie, and Owen came barreling around the corner and ran into their own parents. "It moved! It moved! It moved!" was all he could hear.

Chapter 5

Swamp Thing

"It moved!" Owen repeated.

JT was suddenly very, very glad he hadn't gone with them.

"What moved?" asked Zookeeper Jenny.

"The...the Loup Garou," Evie whimpered. "Owen saw it move, then we all ran away."

"I didn't see it move," said Owen. "Winston saw it move, but it was so scary!"

"I don't know what happened," said Winston. "Sonja shrieked and I was out of there."

"Hmm. I'm sure something moved. Let's go investigate, shall we?" said Zookeeper Jenny. None of the kids followed.

"Oh. I see what must have happened." Zookeeper Jenny bent to pick up a football off the ground.

"The Loup Garou can be scary, so sometimes we like to dress him up for the seasons to make him a little less scary. I brought my supplies over earlier to dress him up for football season. The football must have fallen off the box. I'm sorry it scared you," she said.

The grown-ups smiled and laughed. The kids...not so much. None of them wanted to get too close.

Zookeeper Jenny tried her best to convince the kids that the Loup Garou in the zoo was harmless. She let Baba climb into the exhibit to poke it.

She even tried showing them the collection of silly hats and costumes they used to decorate it.

In the end, though, they had to retrace their steps to leave out of the entrance, because none of the kids wanted to pass the Loup Garou to get to the exit.

Back at the campsite, it was obvious that the grown-ups all thought they were perfectly safe. They kept wanting to sing songs and make s'mores. There even was talk of "going to bed" and "sleeping" which just sounded like plain gibberish.

Who could sleep at a time like this? This was a time for action.

As a scientist, Jaden Toussaint knew that when you have a problem the best place to start is research. Odds were, someone else already had your problem. And odds were, someone else had already written down the solution for you. All you had to do was find it.

JT and Winston flipped through JT's animal encyclopedia to see if there was any information that might help them.

There were lots of facts about wolves, and there were lots of facts about humans, but there were no facts about humans who could turn into wolves or humans who could turn into wolves and sometimes pretend to be plastic statues in zoos.

Sonja, Owen, and Evie couldn't sleep either, so they came to help with the research.

Five heads are better than one, unless five heads are trying to look at the same book all at once. Then the five heads just feel squished.

Jaden Toussaint wished Mama had let him bring more of his animal books. He was right. This was definitely an animal emergency. Luckily, Jaden Toussaint remembered something.

They hadn't had a chance to explore the zoo's animal facts library yet!

Zookeeper Jenny tried to steer them toward the Folktales and Legends section, but JT and his friends knew that arming themselves with real, factual information was the only way to do battle. Stories were not going to help.

What they needed were hardcore facts.

JT passed out their orders.

"Sonja and Owen, learn everything you can about wolves."

"Got it."

"Evie, see if you can find anything about statues coming to life. Real ones," he added. "Not stories."

"On it."

"Winston, looks like the Loup Garou is either Cajun or Canadian. Either way, it speaks French. Can you be in charge of translations?"

"Mais bien sûr!"

"I'll check the fossil record. Let's get to work!"

By morning, they had not found any real information about the Loup Garou. All they had found were a few storybooks from the folktales and legends section and the end of their parents' patience.

Maybe zookeeper Jenny was right. Maybe the Loup Garou was just a story. But if it was just a story, why was it so scary? He'd figure it out. All he needed was a good night's sleep.

Chapter 6

Insomnia

Unfortunately, a good night's sleep was nowhere to be found.

The next night, JT tried to fall asleep in his bed. He couldn't.

Then he tried to go to sleep in Sissy's bed. That was even worse.

After that, Mama tried something extra special. She gave him tea with honey and read him the sweetest, happiest bedtime story they had. It almost worked, but as soon as Mama stopped reading, the Loup Garou crept right back into JT's head.

By the time Jaden Toussaint made it to the rug for circle time the next day, he was really not at his best.

Neither was anyone else.

His teacher, Ms. Bates, was really smart, so she could tell there was a problem. "What's going on? You guys are usually so bright-eyed and bushy-tailed."

Jaden Toussaint snapped to attention. "Did you say bright-eyed?" he asked nervously.

Ms. Bates nodded.

"And bushy-tailed?" Sonja asked, wringing her hands.

Ms. Bates nodded again.

"Ahhh! The Loup Garou!" they all shouted. Evie had struck a karate pose, ready to defend herself. Winston had covered his head with a sweater, and Owen had disappeared to goodness knows where. They were all terrified.

"What's all this about the Loup Garou?" Ms. Bates asked. "Oh, and Owen, you can climb out of that cubby now."

Owen climbed down to join them, and they all told Ms. Bates about their adventure in the zoo.

"It's not even real," said Sonja.

"But it's still so scary," said Evie with a shiver.

"And Sissy isn't even here to help me get back to sleep. Now I may never sleep again," said Jaden Toussaint.

"That does sound like a problem," Miss Bates said. "What could we do to solve it?"

Thanks to his gigantic brain, most ideas just popped right into JT's head without him trying. But the Loup Garou was a toughie. For stubborn problems like this one, there was one surefire way to kick his brain into top gear:

3 MINUTE
STARFISH
DIAMOND SWORD
ALLIGATOR
TURTLE
WOLF

DANCE PARTY
SHAPESHIFTER
Swamp thing
PELICAN
LOOK OUT!
I THINK IT'S... YES!
FAIS DO DO

Then, out of nowhere, it started. That swirly, whirly, zinging feeling he got whenever he was on the verge of a brilliant idea. And just like that, Jaden Toussaint knew what to do.

Chapter 7

Tall Tales

"Well, that was fun, but what are we going to do about the Loup Garou?" Ms. Bates asked.

"We're going to read the story," JT declared.

"What?!?" they all said at once.

"No offense, but that's a terrible idea," said Owen.

"That's not helping, Owen," said Evie. "He's right, though. It's definitely not your best idea."

Jaden Toussaint just smiled. He thought his plan was great, but sometimes it was better to show people than to tell them.

"Raise your hand if you think the Loup Garou is real."

Owen's hand might have wiggled a little bit, but otherwise nobody moved.

"Ok. Now raise your hand if that Loup Garou in the zoo scares you."

Everybody's hand shot up. Even Ms. Bates put hers up a little.

"See? It's not the realness part that's scary. It's the story part that's scary."

"I don't get it," said Sonja. "How does that help us? The Loup Garou story *is* scary."

"I know," JT said.

Pause.

This time he knew. He really, really knew. And he was so excited to share it that he felt like a rocket ship ready to launch.

Unpause.

"The Loup Garou story we know is scary, but if it's a story, that means somebody made it up.

And if somebody else made up a scary one...we could make up not scary ones!"

"That's so crazy..." Winston began. "...that it just might work," Sonja finished.

Jaden Toussaint beamed.

"Well," Ms. Bates said, "we were going to start the day with literacy centers. New Garou stories sound perfect."

Jaden Toussaint's class got down to work making new stories about the Loup Garou. Instead of all the scary stuff, they made stories the way they liked.

Winston's New Garou taught yoga.

Owen's New Garou was a friendly pirate.

Evie's New Garou was an action hero.

And Sonja's New Garou left rainbows wherever she ran.

Jaden Toussaint was the last to take a turn at sharing circle. His New Garou was on a quest to find a good knight's sleep.

And do you know what? That did the trick.

By the end of Jaden Toussaint's story, his New Garou had found a good knight's sleep, and so had everyone else. They slept so long and hard that when Jaden Toussaint woke up and saw Sissy standing over him, he thought it must be a dream.

It wasn't. He had been so distracted by the Loup Garou that he completely forgot she was coming back that day.

Sissy hugged him, and he hugged her right back. "I thought I was tired from Camp Planet Earth, but your whole class is knocked out. What have y'all been doing?"

"I'll tell you the whole story, but can I tell you at the zoo? I think the Loup Garou is going to want to hear it, too."

And that's exactly what they did.

Reading the Loup Garou his new adventure stories was awesome,

but trading adventure stories with Sissy was even better. With stories like these and a sister to share them with, he knew that he would always be the greatest.

EXIT

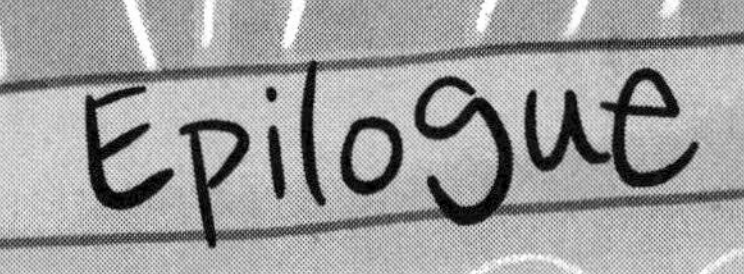

Loup garous definitely just made up stories.

Probably. ~~99%~~ 97.2% sure. Just in case...

How to Stop a Loup Garou from Scaring You

1. Hold a stick and shout, "Ridiculous!"
2. Offer the loup garou s'mores. Prefer beignets?
3. Tell it a happy story

(note: learn to speak loup garou. French + wolf?)

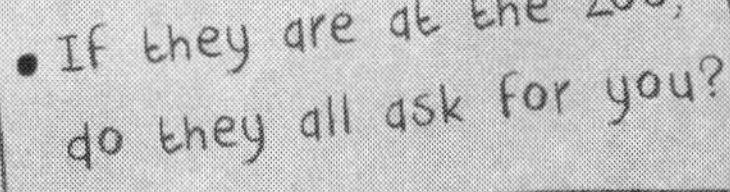

About the Illustrator

Stephanie Parcus loves to create beautiful things. She grew up in Brazil, living off cartoons and bananas, then moved to Italy when she was 10 where she discovered Anime and Manga. Her dream to be a Pokémon trainer and her love for her dog, Fly, led her to become a veterinarian! When she's not on her farm retreat in Italy, she is traveling the world with the human who stole her heart.

About the Author

Marti Dumas is a mama who spends most of her time doing mama things. You know - feeding ducks in parks, constructing Halloween costumes, facilitating heated negotiations, reading aloud, throwing raw vegetables on a plate and calling it dinner, and shouting, "Watch out!" whenever there are dog piles on the walk to school. Sometimes she writes, but only very occasionally and in the early morning.

You can find her at:
www.MartiDumasBooks.com

JADEN TOUSSAINT, THE GREATEST

EPISODE 1: THE QUEST FOR SCREEN TIME

Written by Marti Dumas

Illustrated by Marie Muravski

JALA
and the
WOLVES

by Marti Dumas

For crafts, recipes, and more, visit:

www.MartiDumasBooks.com

Authors love reviews.
We eat them up like
pineapples for breakfast.

Yum!